Ny Aina Nambinintsoanirina RAVOAVY

IRAT 200 maize cultivation: experimentation with new fertilizers

Ny Aina Nambinintsoanirina RAVOAVY

IRAT 200 maize cultivation: experimentation with new fertilizers

ScienciaScripts

Imprint
Any brand names and product names mentioned in this book are subject to trademark, brand or patent protection and are trademarks or registered trademarks of their respective holders. The use of brand names, product names, common names, trade names, product descriptions etc. even without a particular marking in this work is in no way to be construed to mean that such names may be regarded as unrestricted in respect of trademark and brand protection legislation and could thus be used by anyone.

Cover image: www.ingimage.com

This book is a translation from the original published under ISBN 978-620-6-71205-3.

Publisher:
Sciencia Scripts
is a trademark of
Dodo Books Indian Ocean Ltd. and OmniScriptum S.R.L publishing group

120 High Road, East Finchley, London, N2 9ED, United Kingdom
Str. Armeneasca 28/1, office 1, Chisinau MD-2012, Republic of Moldova, Europe
Printed at: see last page
ISBN: 978-620-7-61556-8

"I give you thanks for so many wonders: marvellous are your works"

Psalms 139:14

ACKNOWLEDGEMENTS

A number of people have contributed to the production of this memoir. We are deeply and sincerely grateful to all of them.Special thanks go to Mr ANDRIAMANIRAKA Jaona Harilala, Doctor of Agronomic Sciences and Research Teacher at ESSA (Ecole Supérieure des Sciences Agronomiques), Head of the Agriculture Department at ESSA, for having done us the honour of chairing the jury for this dissertation and for all the attention you have given it.Our thanks go to the AGRIVET company, without which this thesis would not have been possible, and more particularly to Mr HERINDRANOVONA Augustin, AGRIVET Research and Development Manager, for agreeing to supervise us professionally.

Our gratitude goes to Mr. RAKOTO Benjamin, Research Lecturer at ESSA, for having supervised us throughout this work. For ***all the*** *advice and criticism that helped us a great deal in finalising this manuscript. We would also like to thank Ms RAMANANKAJA Landiarimisa, PhD in Agriculture, Research Teacher at ESSA, for agreeing to sit on the jury and for examining this dissertation.We would like to thank Mrs LALANEKENARISOA Nenée, Head of Operations at the Centre de Formation et d'Application du Machinisme Agricole or CFAMA Antsirabe, for her help, advice and warm welcome.*

To all the staff of AGRIVET, To all the staff of CFAMA Antsirabe, To my family in particular: Dada sy Neny, Malala, Kokoly and Lahatra. To my friends and colleagues at ESSA To all ESSA staff, especially those in the Agriculture DepartmentOur thanks go to all those who, from near or far, have made a contribution to the success of this project. award for this brief. Thanks to all

SUMMARY

With its various uses, maize is one of the most important crops in Madagascar. However, maize yields among farmers are still low. Fertilisation plays a vital role in increasing plant yields. The AGRIVET company has therefore conducted an experiment on maize fertilisation in the Vakinankaratra region, a maize-growing area. The aim of the study was to determine the effects of different fertilisations on maize. Different combinations and doses of fertilisers with covers were tested on IRAT 200 maize, a variety already adopted by farmers. Two types of phosphate fertiliser were used, TSP and calcite. For nitrogen (N), three different doses were applied: 0Kg/Ha N, 60Kg/Ha N and 120Kg/Ha N. Three doses of cover were used: 0%, 30% and 60%. 18 treatments were tested with three replications. After these experiments, three treatments (C1 + N1 + P1, C1 + N2 + P1, and C2 + N2 + P1) obtained the highest yield, ranging from 4T/Ha to 5T/Ha. The results showed that TSP is very effective in achieving high yields. However, calcite is also effective from an agronomic point of view. Calcite made up for the lack of nitrogen in treatments where no nitrogen was applied. An increasing dose of nitrogen on the maize crop significantly increases grain maize yields. Despite the contribution of organic matter by the cover crops, they did not affect yield.

Key words: AGRIVET, urea, phosphorus, fertilisation, plant cover.

INTRODUCTION

Along with rice and wheat, maize (zea mays) is one of the three most widely grown grasses in the world. World production is estimated at 943 million tonnes for the 2013-2014 season, compared with 863 million tonnes in 2012/2013 (CIC, 2014 and Limagrain, 2013). It occupies a global area of almost 180 million hectares, with production rising steadily (agpm.com, 2012).Maize plays an important role in Madagascar's economy. National production, estimated at 380,848 tonnes in the 2012/2013 season (INSTAT and FAO, 2013), from a total cultivated area of 200,000 hectares, makes a significant contribution to meeting the population's basic food needs, particularly during the lean season (MAEP UPDR - océan consultant, 2004).

The main production areas are Lac Alaotra, the Central Highlands, the Middle West and the South West, accounting for over 97% of total national production (INSTAT, 2010). Essentially considered as a staple food for the population of the South after rice and manioc, maize is also used in animal feed (feed component, poultry feed) and in industry (brewing, soap production, etc.) (CIRAD and GRET, 2006).

Given its importance and its various uses, a good production with a high yield of maize crop is useful not only for animal feed and industry but also for the population. Several factors such as meeting water requirements, soil type, genetic potential, pest problems (ANDRIAMAMPANDRY., 1990) ... influence the productivity of all crops. Fertilisation is one of these yield factors. Good fertilisation leads to good productivity.

With this in mind, AGRIVET, in collaboration with the International Atomic Energy Agency (IAEA) and the radioisotope laboratory (LRI), is carrying out research and experiments on the mineral fertilisation of maize in maize-growing regions of Madagascar over a five-year period. Experiments using different doses and sources of phosphate and nitrogen fertilisation are being carried out in the Vakinanakaratra region, which is one of the main maize-producing areas.The aim of this experiment is to find out the effects of combining different fertilisers on maize yields. In addition, we want to know the effect of using Calcite, a natural phosphorus, a new product on the market. In this study, IRAT 200 is the variety being tested. IRAT 200 is one of the main varieties adopted by farmers in Madagascar, along with CIRAD 412, MEVA and 'Mailaka. (INRA, 2011 and FARE Y, 2004)

Hence the title of this work **"Effect of different fertiliser combinations and doses** on maize cultivation. The case of the IRAT 200 variety in the Vakinankaratra region".

So the question is: **"Which of the various fertiliser combinations for IRAT 200 maize gives the best yields that can be distributed to farmers?**

A number of hypotheses have been put forward to answer this question:

- **First hypothesis**: among the treatments tested, there is a combination giving a very high yield compared with others.
- **Second hypothesis**: calcite has a positive agronomic effect, i.e. on soil properties and grain maize yields, compared with Triple Super Phosphate or TSP.

This study will therefore analyse the effects of combining different forms and doses.

It is divided into three main sections:

- Firstly, the materials and methodologies used,
- monitoring results
- and at the end the discussions and recommendations.

I- MATERIALS AND METHODS

1- Study site

For this study, the Vakinankaratra region was chosen as the experimental site, as it is one of the main maize-producing regions in Madagascar. Maize is one of the main crops grown in Vakinankaratra, along with rice, manioc, sweet potatoes, beans and potatoes. It is grown almost everywhere in the region. Collaboration has been undertaken with the CFAMA agricultural machinery training and application centre in Antsirabe to set up experimental plots.

Figure 1: Location map of the study area

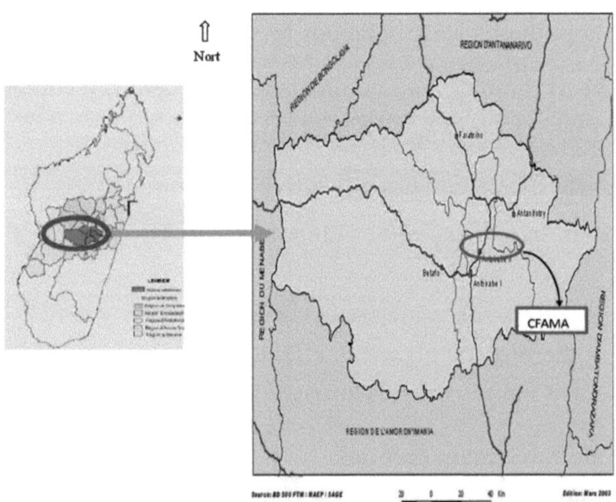

Source: UPDR (Rural Development Policy Unit) 2003

1-1- Climate of the study area

The Antsirabe area has a tropical climate at an altitude of over 900 metres, with frosts between June and August. The average annual temperature is less than or equal to 20°C (UPDR, 2003).

Figure 2: Average monthly rainfall in Antsirabe 2009 to 2014

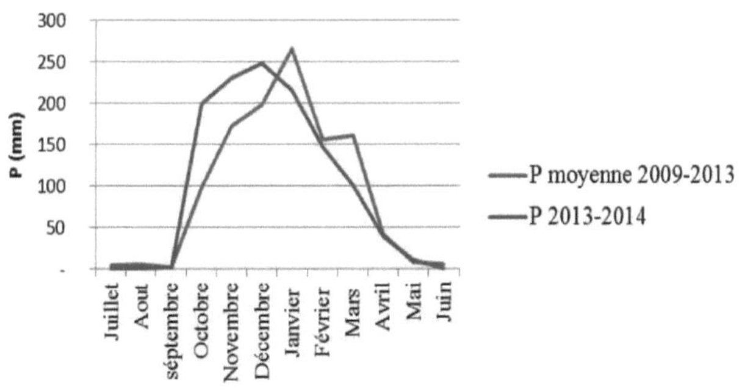

Source: author

According to this figure, rainfall for the 2013-2014 season shows a different pattern to that of the last five years. The average rainfall for the last five years peaked in January, whereas for the 2013-2014 season it peaked in December. In November and December, rainfall for the season was well above the average for the last five years, exceeding 100 mm. The water requirements of maize during these two months were therefore met, as maize needs 100 mm/month of water during its crop cycle (FARE Y, 2004). The average temperature of the study area over the last five years and that of 2013-2014 are more or less the same. Between November and February the temperature fluctuates around 19 to 20°C. The November temperature of 20°C is ideal for germinating seedlings. Maize needs an optimum temperature of 19°C during its growth (Mémento, 2006), so the average monthly temperature at CFAMA is favourable for its development.

Figure 3: Average monthly temperature in Antsirabe from 2009 to 2014 1-2 Soil and vegetation

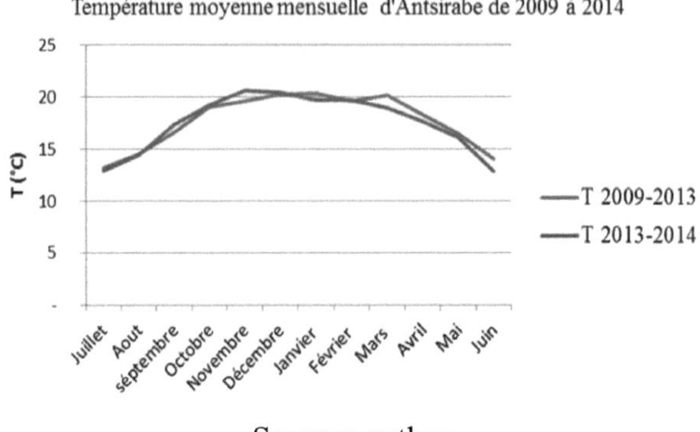

Source: author

The Vakinankaratra Region is characterised by the dominance of two types of soil (UPDR, 2003) Ferralitic soils cover a large part of the region. They can be used for maize, manioc, potatoes and arboriculture. The alluvial soils that make up the lowlands are used to grow off-season crops in addition to rice. The Vakinankaratra region is characterised by a small area of primary forest. Degradation is such that only a few remnants of forest remain in the region. In the low-lying areas, there are rush marshes and remnants of gallery forests which are disappearing. (UPDR, 2003).

1-3 Description of the experimental site

The experimental site is located within the CFAMA cultivation areas with a total surface area of approximately 16.65 ares. At the CFAMA study site, almost all the land is planted with food crops such as several varieties of rice (FOFIFA 154, X134, etc.) and maize of the MEVA variety. There are also soya and vegetable crops. With regard to the cropping history of the experimental field: CFAMA practised a biennial rotation of soya and rainfed maize-rice between 2008 and 2013.

2008-2009: Soya;

2009-2010: Pre-basic seeds for rainfed rice with isolation between varieties, maize; 2010-2011: Soya;

2011-2012: Pre-basic seeds for rainfed rice with isolation between varieties, maize; 2012-2013: Soya beans

2- Material plants

The plant material is maize or zea mays (Cf.: Appendix I). The variety used in this experiment is IRAT-200, which originates from Côte d'Ivoire and has a cropping cycle of 105-120 days.

The seeds are horny to toothed and orange-yellow in colour. The variety has slightly curved leaves, a weak zigzag stem and a cylindrical-conical spike shape. The plant is around 2.20 m to 2.50 m tall.

Figure 4: IRAT 200 maize grains

It is well adapted to all of Madagascar's agro-climatic conditions, especially in coastal regions (except the east coast), with yields of between 5.4 and 6.6 tonnes per hectare (MINAGRI et al, 2010).

Table 1: Characteristics of IRAT 200

Varieties	Types	Type of grain	Cycle	Colour
IRAT 200	Composite	Toothed	105 to 120 days	Orange yellow

Source: MINAGRI et al, 2010

3- The systems studied

During the experiment, the parameters to be considered in the experimental set-up were:

- Crop residue coverage: there is zero coverage (C0), 30% coverage (C1) and 60% coverage (C2).

The cover used was dry maize matter. The rate of cover applied was 2.5 tonnes/Ha, as a maize crop produces around 2-3T/Ha of dry matter (RASOAMAHARO, 2008).

- Nitrogen dose: with zero input (N0), an input of 60KG/Ha (N1) and an input of 120KG/IIa (N2). Nitrogen was supplied in the form of urea.

- Phosphorus input: either use of TSP (P1) or use of phosphate which is calcite ("P2).

The experiment therefore included 18 systems to be studied.

Let's look at the chemical composition of calcite and TSP :

- Composition of calcite :

$P O_{25}$ > 20 % CaO= 40 %

Total N > 0.02

$K O_2$= 0, 3 %

SiO_2 = 0.5

MgO= 0, 3 %

Humus > 1

Calcite contains 20% phosphorus and a large quantity of CaO of approximately 40%.

- Composition of the TSP :

Source: New Caledonia Chamber of Agriculture, 2000

Nom anglais	Triple Superphosphate (TSP)			
Autre nom français	Phosphate monocalcique (composant essentiel)			
Provenance	Sico (Belgique) et Bush Int. (NZ).			
Formule chimique	$Ca(H_2PO_4)_2$, H_2O			
Analyse	**N**	**P_2O_5 (P)**	**K_2O**	**CaO**
	0	**46 %** (20 %) (91% soluble dans l' eau)	**0**	**15 %**
	Autres : S : 1,3 %			
Solubilité	Assez soluble dans l' eau (18 g/L).			
Humidité	6,20 % max.			
Granulométrie	< 2 mm : 1 % 2-4 mm : 85 % > 4 mm : 14 %			
Densité	0,85-0,95 t/m^3			
Hygroscopicité	Très faible (au delà de 95 % d' humidité relative à 30 °C).			
Stockage / précautions	Garder le sac fermé avant utilisation. Stocker en dehors de l' action directe du soleil. Se laver les mains après utilisation.			
Compatibilité	Mélange toujours possible : - ammonitrate, nitrate de potassium, sulfate de potassium, 0-32-16, 13-13-21, 16-4-8, fumier, engrais organiques. Mélange possible au moment de l' emploi : - urée, MKP, 17-17-17, gypse. Ne jamais mélanger avant emploi : - nitrate de calcium, hyperphosphate, chaux.			
Effet sur le pH	Peu d' effet sur le pH du sol.			

TSP contains around 46% phosphorus and 15% CaO.

4- Experimental

The experimental set-up consists of 54 elementary plots 5 m long and 4 m wide, divided into 3 blocks. Each plot is separated by a 1 m wide aisle and a 2 m aisle between the blocks. The system is a Fischer block with 3 replicates.

Figure 5: The experimental set-up

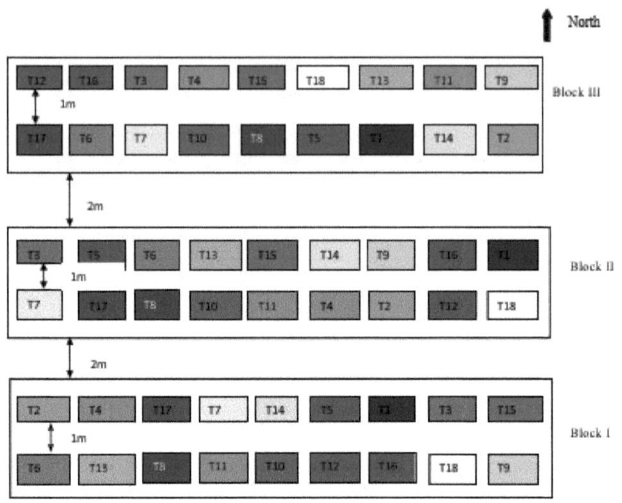

Source: author

T1: C0 + N0 + P1

T2 : C0 + N0 + P2

T3 : C0 + N1+ P1

T4 : C0 + N1+ P2

T5 : C0 + N2+ P1

T6 : C0 + N2+ P2

T7 : C1 + N0 + P1

T8 : C1 + N0+ P2

T9 : C1+ N1+ P1

T10 : C1+ N1+ P2

T11 : C1+ N2+ P1

T12 : C1+ N2+ P2

T13 : C2 + N0+ P1

T14 : C2 + N0+ P2

T15 : C2 + N1+ P1

T16 : C2 + N1+ P2

T17 : C2 + N2+ P1

T18 : C2 + N2+ P2

With :

C0: zero coverage C1: 30% coverage C2: 60% coverage

N0: without nitrogen N1: with 60Kg/Ha of nitrogen N2: with 120Kg/Ha

P1: TSP

P2: calcite

Note: Treatment T18 was eliminated due to the low rate of seedling emergence on the elementary plots. For the results, therefore, we treated 17 treatments.

5- The behaviour of experimentation

5- 1 Preparing the soil and setting up the basic plots :

Ploughing: ploughing was carried out using a power tiller for all plots, followed by spraying and manual ploughing. We ploughed 3 days before sowing, i.e. on 27 November 2013, to a depth of 30 cm. We then removed plant residues from the plots. One day before sowing, we crumbled the clods manually.Plotting: for the plotting, we staked out block by block and then elementary plots. by individual plot using stakes.

5- 2 Sowing and spreading fertiliser

Sowing took place on 30 November 2013 after the elementary plots had been laid out to a depth of around 4-5 cm. With a density of 40,000 plants/Ha, i.e. a spacing of 50cm *40 cm, sowing was done manually with one seed per stake. The density adopted was based on the dimensions of the elementary plots (5m * 4m).Fertiliser was applied at the same time as sowing. Fertilisers were applied manually to the seed rows and to each plot. Potassium sulphate was applied at a rate of 10Kg/Ha K for all plots. For phosphorus at 20Kg/Ha of P, we applied TSP at a dose of 100Kg/Ha on 27 elementary plots and 232kg/Ha of calcite on the other 27 plots. Nitrogen application was split in two, with 1/3 applied at sowing. Of the 54 elementary plots in the three blocks, 18 plots received 130kg/Ha of urea, 18 plots received 260kg/Ha and the remaining 18 plots received no nitrogen.

5- 3 Interviews :

Two weedings were carried out during the plant cycle, the first on D_{17} after sowing and the second on D_{49} after sowing. 2/3 of the nitrogen was applied on D_{30} .An insecticide treatment with liquid pyruban, a member of the organophosphate family, was carried out on D_{17} after sowing at a dose of 2.5l/Ha diluted with 300l/Ha of water. On D_{44} , agrimethrin, a foliar insecticide, was used at a dose of 250ml/Ha to treat foliar insects. To control corn earworm, use gazidim at a dose of 1l/Ha diluted with 250l/Ha of water on D_{57} .

5- 4 Observation of the phenological state :

After planting. Climatic observations were carried out using a daily rainfall record and phenological observations (1.2 m x 2 lines in the centre of the plot):

- Lift rate ;
- Flowering time ;
- Time to heading ;
- Ripening time ; 5- 5 Harvesting :

It took place on 1[er] May 2014, D_{151} and was done manually with labour. Harvesting was done by elementary plot by elementary plot starting with block I, then block II and block III.

5- 6 Post-harvest operations :

After harvesting, the weighing, drying and counting operations began. The cobs from the 8 central vines were dried for three days before being weighed. After weighing, the ears were dehulled, 100 grains were counted and the average number of grains per ear was determined. Moisture tests were carried out on a sample of 100 maize kernels for each treatment in the three blocks.

5 -7 Activity timetable

The activities undertaken during the experiment are summarised in this section:

Figure 6: Timeline of activities

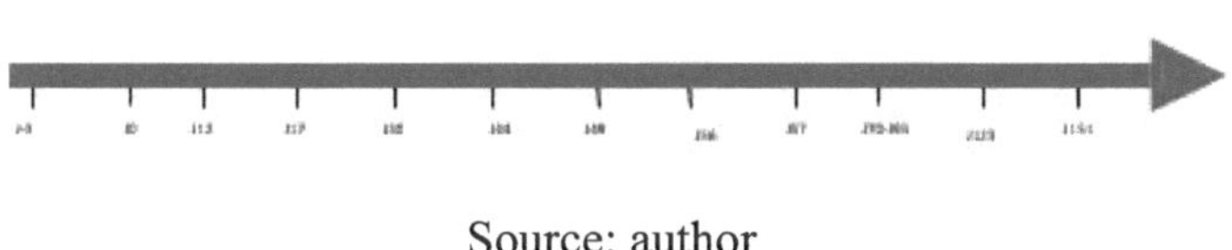

Source: author

With : D_{-3} : ploughing D_0 : day of sowing

D_{12} : observation of emergence rate

D_{17} : first weeding and phytosanitary treatment (pyruban) D_{30} : 2/3 urea application and height measurement

J_{44} : plant protection treatment (agrimethrin) $J_{56:}$ height measurement

D_{57} : phytosanitary treatment (gazidim)

D -D_{7093} : observation of flowering and heading times

J_{123} : measuring height and cutting stems J_{151} : harvesting

6- Parameters measured and observed

- Emergence rate: the emergence rate was measured on D_{13} after sowing. This was done by counting the number of seedlings that had emerged in each plot.
- Height: three height measurements were taken on the 8 central feet on D_{30} , D_{56} and the day before the stems were cut on D_{123} .
- Flowering, heading and ripening: for these parameters we observed the date of appearance of male and female flowers and mature plants.
- Rainfall: using meteorological data (Ampandrianomby station)

• Yield and yield components: to assess yield, we have used this scheme for calculating maize yield:

Figure 7: Maize yield calculation

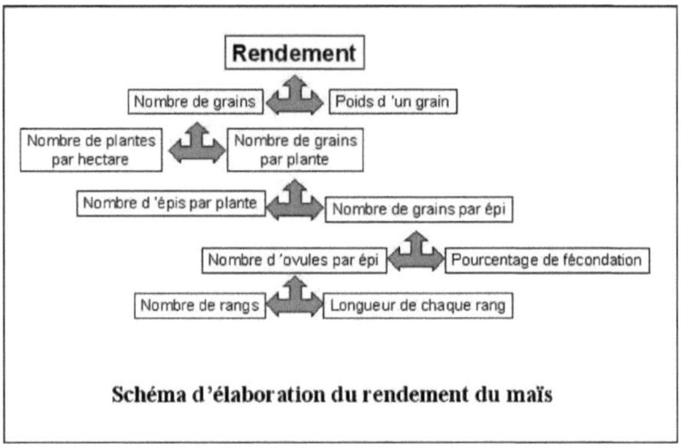

Rendement

Nombre de grains | Poids d 'un grain

Nombre de plantes par hectare | Nombre de grains par plante

Nombre d 'épis par plante | Nombre de grains par épi

Nombre d 'ovules par épi | Pourcentage de fécondation

Nombre de rangs | Longueur de chaque rang

Schéma d'élaboration du rendement du maïs

Source: CIRAD and GRET, 2006

According to this diagram of maize yield development, yield is assessed by the following formula :

RDT (T/Ha)= weight of one grain x number of plants/Ha x number of spikes/plant x number of spikes/plant x number of spikes/plant of grain/ ears

With RDT: yield

During weighing, we weighed 100 grains for each treatment in the three blocks. The weight of a grain was obtained using the weight of 100 grains.

The number of spikes per plant was obtained by counting the spikes from the 8 plants during harvest. for each of the plots. The number of kernels per ear was assessed using the total weight of the kernels (weighed after

7- Processing data

The data obtained were processed using the experimental statistical method with XL STAT 2008 software. For the treatments we used analysis of variance (ANOVA) with level 4 interaction and the coefficient of determination R^2 to analyse the influence of the treatments on the parameters measured. Linear regression was used to determine the influence of yield components on yield. Comparisons of means between blocks and treatments and between treatments and measured parameters were made using Fischer pairwise comparisons at the 0.05 significance level.

II- RESULTS

1- Performance

1- 1 Influence of yield components on yield: modelling yield as a function of its components

Table 2: Correlation matrix between yield and its components

Variables	no. of grains per ears of corn	number of ears	weight 1000 grains	Rdt
Number of grains per ear	1	-0,074	0,425	0,781
number of ears	-0,074	1	0,018	0,362
weight 1000 grains	0,425	0,018	1	0,747
Rdt	0,781	0,362	0,747	1

Source: author

Table 3: Coefficients of determination R^2 between yield and its components

Variables	no. of grains per ears of corn	number of ears	weight 1000 grains	Rdt
number of grains per ear	1	0,006	0,181	0,610
number of ears	0,006	1	0,000	0,131
weight 1000 grains	0,181	0,000	1	0,558
Rdt	0,610	0,131	0,558	1

Source: author

Among the yield components, the weight of 1000 grains and the number of grains per ear are the most telling variables. R^2 between yield and number of grains per ear is equal to 0.61, which means that 61% of yield variability is defined by the number of grains per ear.

1- 2 effects of treatments on yield components

• Number of grains per ear

Figure 8: Number of grains per ear for cach treatment

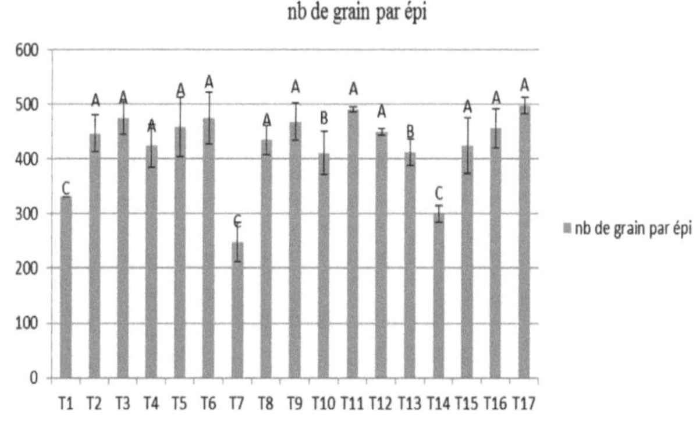

Source: author

According to the ISS variance table, the factor studied (combination of fertiliser and cover) has a significant effect on the number of grains per ear because the probability Pr > F (0.0001) is well below 0.05. According to the coefficient of determination $R^2 = 0.74$, 74% of the variability in the number of ears is explained by the treatment. In general, the number of grains per ear varies from 248 to 491, with treatment T7 having the lowest number of grains per ear and treatment T11 the highest. From the figure we can classify the treatments into three groups according to the number of grains per ear: Group A: made up of treatments giving the highest number of grains per ear, varying between 424 and 491: T11, T17, T6, T3, T9, T5, T12, T16, T2, T8, T15, and T4.

- Group B: 332 to 412 seeds per spike with: T13 and T10.

- Group C: with the lowest number of grains per ear between 248 and 300 with T1, T14 and T7.

- Number of spikes per plant

Figure 9: Total number of ears for each treatment

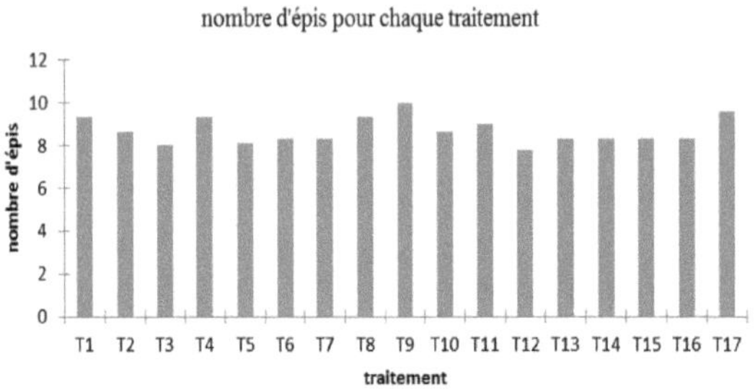

nombre d'épis pour chaque traitement

traitement	T1	T2	T3	T4	T5	T6	T7	T8	T9	T10	T11	T12	T13	T14	T15	T16	T17

Source: author

In these figures for the treatments, the number of spikes varies from 8 to 10 for the 8 central plants, so the number of spikes per plant is 1 to 1.2 on average because there are plants bearing 2 to 3 spikes. The plots with treatment T9 gave the highest number of ears, around 10. According to the ISS analysis of variance, the probability Pr > F (0.65) is greater than 0.05. This means that the treatment had no effect on the number of ears per plant or on the number of ears on the 8 central plants. Consequently, there is no specific group for each of the treatments.

- Weight of 1000 grains and one grain

In the graph below, the weight of 1000 grains of IRAT 200 maize ranges from 180 g to 22 0g on average and the average weight of one grain varies from 0.18 g to 0.22 g. The weight of 1000 grains and one grain for treatments T13 and T9 is the highest and that of T1, T12 and T14 is the lowest. There is no distinct division of treatments according to the weight of 100 grains and one grain because the difference between the weight values is not significant. In fact, the combination of forms and doses of fertiliser and cover had no influence on the weight of 1000 grains of the plant with a probability Pr > F (0.7) greater than 0.05.

Figure 10: Weight of 1000 grains in grams for each treatment 1- 3 Grain maize yield

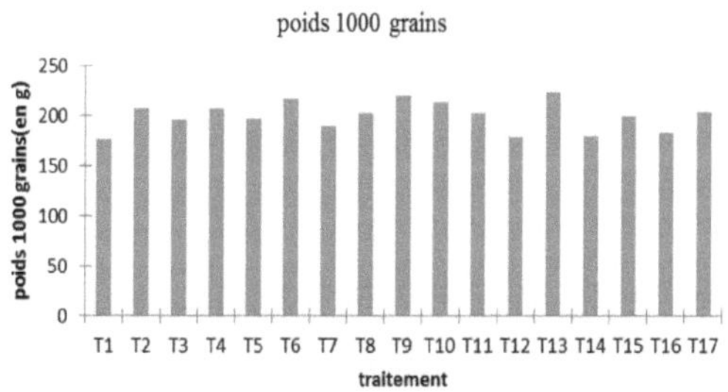

Source: author

According to the ANOVA table (see Appendix II), the interaction between yield and treatments had a significant effect overall, as Pr > F 0.016 is less than 0.05 (significance level). For the coefficient of determination R^2 , on the model studied R^2 =0.59of the results are explained by the factor studied and the block factor. The interaction between yield and the factor studied had a significant effect, so we proceeded to the comparison of means test based on this interaction.

Figure 11: Grain maize yield for each treatment

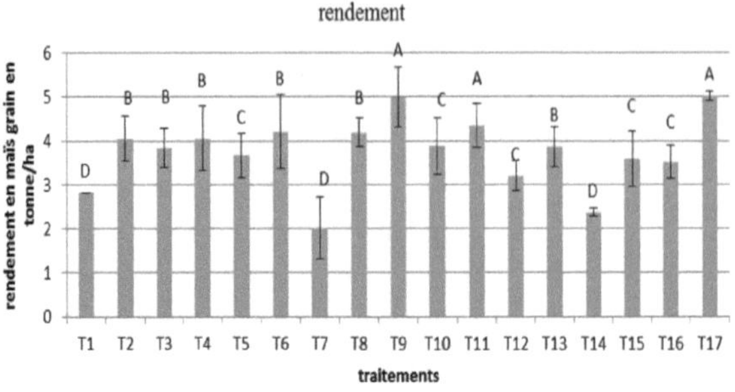

Source: author

Yields for all plots in the three blocks generally ranged from 2 to 5 T/Ha. Based on the yields obtained, we can classify four distinct groups:

- Group A: represents treatments with the highest yields between 4.3T and 4.9 T/Ha. This group is made up of : T11, T17 and T9.
- Group B: T2, T4, T6, T8, T3 and T13 with yields of between 3.7T and 4.2T/Ha.
- Group C: with a yield of 3.2 to 3.7T/Ha with : T5, T15, T16, T12 and T10.
- Group D: yield less than 3T/Ha: T7, T14 and T1.

2- Phenological status

2-1 Emergence rate

According to the ANOVA table of the lift rate on treatments :

- The coefficient of determination R^2 =81% means that the emergence rate depends on the factor studied, which is the combination of forms and doses of fertiliser and cover on the IRAT 200 maize crop. The relationship between the two variables is strong, with 81% of the variation in emergence rate depending on the treatments.

- According to the analysis of variance and the standard ISS analysis, the probability obtained Pr > F (cf. appendix I) is< 0.0001, which is below the significant threshold of 0.05. Overall, the factor studied had a significant effect on the lift rate. This explains the division into three groups of treatments according to their emergence rate. According to the ANOVA of the estimated means (see appendix II), the treatments can be divided into three groups:

- Group A: with the highest lift rate of over 65%, this group comprises : T1, T2, T7, T8, T13 and T14.
- Group B: T15, T16, T3 and T9 with lift rates ranging from 50 to 65%.
- Group C: T6, T5, T11, T10, T12, T4 and T17 with the lowest lift rates of between 35% and 50%.

Figure 12: Emergence rate for each treatment

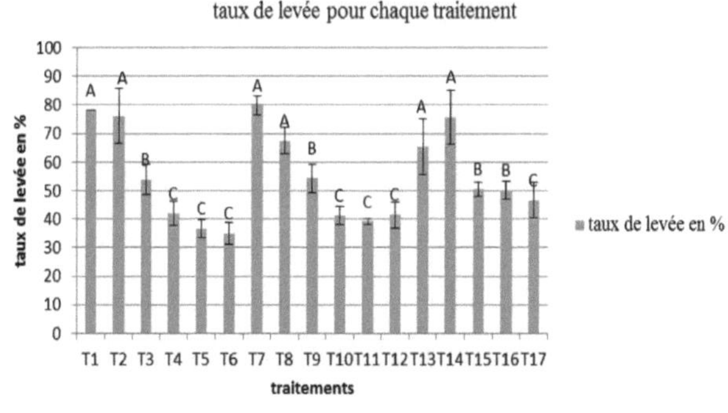

Source: author

Two to three weeks after sowing, the average emergence rate was 55%. The graph above shows the emergence rate of maize plants for each fertiliser and cover combination.

2-2 Observation of flowering time

Flowering or the appearance of the male inflorescence began on 14 February, i.e. D_{76} after sowing, i.e. 11 weeks after sowing. On 15 February or D_{77} of the 8 central plants in the plots, some had flowered halfway (4 plants), others 25% (2 plants) and others had not yet flowered. Between D_{76} and D_{86} 24 February, all the central plants in the plots had flowered.

2-3 Observation of heading

The appearance of the female inflorescence in the plots began about a week after the appearance of the male inflorescences at around D93 after sowing.

2-4 Observation of ripening

Plant maturation began on D_{115} , i.e. 115 days after sowing. We found that it was the plots with the N0 treatment that were mature on D_{115} . On D_{121} , all the plots with N0 were mature. The other plots ripened between D_{121} and D_{128} and the stems were cut. It should be noted that the N2 plots ripened later than the N0 plots.

3- Height

Three measurements of crop height were taken during the growing cycle. The first was taken around D_{30} , the second around D_{56} and the last on D_{123} after sowing.

Figure 13: Heights of IRAT 200

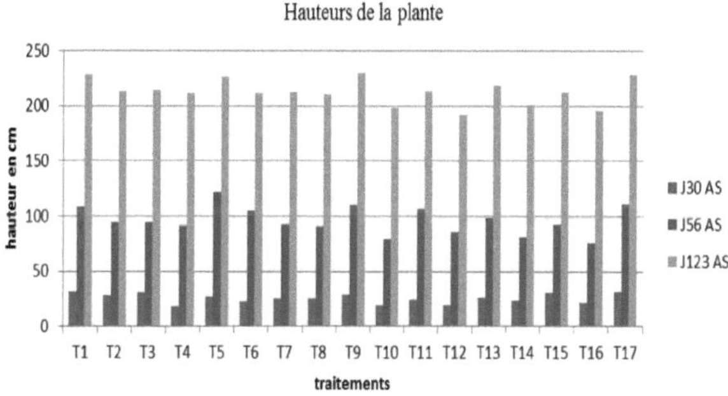

Source: author

According to this figure, the height varied from 18 cm to 31 cm one month after sowing. Approximately two months after sowing, the height ranged from 73 cm to 110 cm. Before the stems were cut, i.e. around D123 after sowing, the height varied from 193 cm to 228 cm. For the final height before maize harvest, treatments T9 and T17 have a very high height compared to other treatments such as T12 and T16 as shown in the figure below.

Figure 14: Graph of average heights for each treatment before harvest (H3)

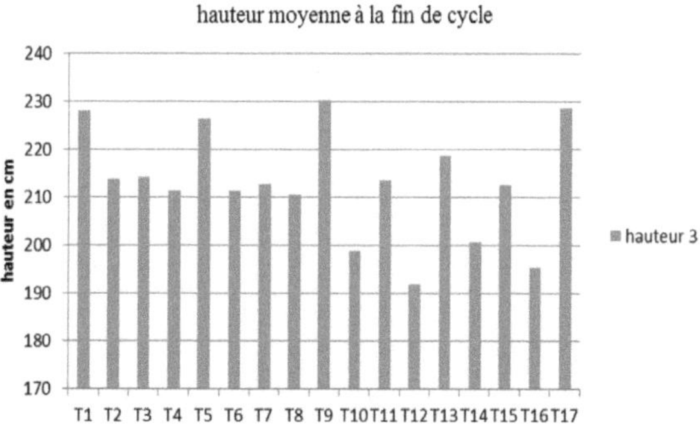

Source: author

Table 4: Correlation matrix between yield and final plant height

Variables	Rdt	height 3
Rdt	1	0,382
height 3	0,382	1
	Source: author	

The coefficient of determination (R^2) for yield and height was 0.146. Only 14.6% of the variability in yield was explained by plant height.

Height and treatment

Height 1: J_{30} AS: R^2 =0.46 46% of the variability in maize height 30 days after sowing is explained by the treatments. Overall, the interaction between the factors did not have a significant effect because Pr > F (0.945) is well above 0.05. But the phosphorus factor did have a significant effect. However, the phosphorus factor did have a significant effect, and we proceeded with a pairwise comparison between P1 and P2.

Table 5: Estimated average heights J_{30} AS using P1 and P2

Modality	Estimated mean	Groups P1	28,409
	A		
P2	22,847	B	

Source: author

There was a significant difference between the two averages. The height obtained by P1 is 28.4 cm and by P2 22.8 cm.

Height 2 days$_{56}$ AS: R2=0.67: 67% of the variability in maize height 56 days after sowing is explained by the treatments.

Overall, the interaction between height 2 (plant height at D56 after sowing) and the treatments had no significant effect because Pr > F (0.237) is well above 0.05. However, the phosphorus and nitrogen factors did have a significant effect, and we proceeded with a pairwise comparison between P1 and P2.

Table 6: Estimated average heights J_{56} AS using N0, N1 and N2

Modality	Estimated average		Groups	
N2	106,485	A		
N1	94,083			B
N0	90,444			B

Source: author

The estimated mean for N2 is 106.5 cm, for N1 94 cm and for N0 90 cm. There was no significant difference between the estimated means for N0 and N1. However, these averages obtained with N0 and N1 are different from that of N2.

Table 7: Estimated average heights J_{56} AS using P1 and P2

Modality	Average estimated		Groups
P1	103,713	A	
P2	90,296		B

Source: author

The estimated average plant height was 103.71 cm for P1 and 90.3 cm for P2. The two averages are significantly different.

Height 3: final height R^2 =0.65

65% of the variability in maize height at the end of the cycle is explained by the treatments. Overall, the interaction between the factors did not have a significant effect, as Pr > F (0.188) is above the 0.05 significance level. However, the phosphorus factor did have a significant effect, and we proceeded with the pairwise comparison between P1 and P2.

Table 8: Estimated average heights J_{123} AS using P1 and P2

Modality	Estimated average	Groups
P1 P2	220,299A 206,940	 B
	Source: author	

The pair-wise comparison between P1 and P2 showed that the estimated mean using P1 is higher than that using P2.

III-DISCUSSIONS

1- Effect of the factor studied on grain maize yield

Overall, the factor studied, which is the combination of different forms and doses of fertiliser, influences yield. With the difference in yield obtained by using the different combinations, the treatments were classified into four groups. Group A had very high yields compared with the three groups (B, C and D). As for the high-yielding treatments in Group A, T9 (C1N1P1), T11 (C1N2P1) and T17 (C2N1P1), the conditions for good production in these treatments were met. Nitrogen is used, the essential factor for high yields. Phosphate mineral supplementation using TSP, which is very useful because it suddenly enters the soil food pool and available directly to the plant (RANDRIAMANANANDRO A, 2010). Presence of plant cover, a source of organic matter. Even if the interaction between cover and yield did not have a significant effect, the use of cover helps to increase the supply of nutrients to maize and is of interest for safeguarding the future of the soil, given that maize is known to impoverish the soil (GROS A, 1962).Nitrogen is an essential factor in growth and yield (FAO, 1980). It has a shock effect on vegetation. A plant that is well supplied with nitrogen grows quickly and takes on a beautiful dark green colour due to the abundance of chlorophyll. Since photosynthesis takes place in the green parts, which are full of chlorophyll, we can say that it is in these parts that the yield is obtained. What's more, the chlorophyll that underpins the fundamental operation of photosynthesis is a nitrogenous material. Nitrogen is therefore the determining factor in yields and the pivot of fertilisation (GROS A., 1962), which explains why plots treated without nitrogen (T1, T14 and T7) have low yields compared with plots treated with nitrogen (T9, T17, T11). Nitrogen is the key element in maize cultivation; it is necessary for good plant development and allows normal cob formation (FARE.Y, 2004) as well as an increase in the number of grains per cob (ANDRIAMAMPANDRY., 1990).During field observations, we noticed signs of nitrogen deficiency in some plots in the three blocks towards the appearance of male inflorescences. The deficient plots in block I are the plots treated with: T13, T7 and T1. In block II: T1, T13 and T8. In block III: T1 and T2.A yellowish-green discolouration of the leaves begins_ towards the oldest leaves. The leaves become discoloured around the edges and along the veins.

According to several authors, N deficiency will produce a characteristic V-shaped yellowing on the leaves of maize and will produce a more general pale green or yellow discolouration that is less pronounced as N is translocated to the faster-growing parts of the plant, with the lower leaves being the first to become deficient. Nitrogen deficiency in maize results in the absence of an ear on the stalk, or the stalk bears an ear without grain. The weight of the ear is low, leading to a reduction in the total weight of the kernels, which plays a part in assessing yield (ANDRIAMAMPANDRY., 1990).Deficient plots are those with N0, i.e. no nitrogen input. Maize responds well to nitrogen, and nitrogen deficiency is the second most important constraint on tropical maize production after drought (FAO, 2002). Nitrogen is an important factor in maize fertilisation. A lack of nitrogen can lead to a drop in yield (Thibaudeau S, 2006). This confirms the effectiveness of calcite, as we have seen above: even without nitrogen, production is still good.For groups D, which gave the lowest returns : The application ofT1, T14 and T7 gave the lowest yield (less than 3T/Ha) for the whole study during the experiment. Even though the blankets used in these three treatments were of different doses, this had no effect on yield. However, in terms of the type of phosphate fertiliser used, two of the three treatments were treated with TSP.Mineral phosphate alone is not enough to obtain good yields, even in high doses to ensure maize nutrition (RASOAMAHARO, 2008). TSP contains no other nutrients apart from phosphorus and Ca0 (in small quantities compared with the Ca0 contained in calcite). Phosphate fertilisers combined with organo-mineral amendments or nitrogen must be added to the soil to improve production.For T14, grain maize yields were good, but because of the attack by corn earworms and cutworms on the plot treated in block I, yields were low. The total kernel weight for T14 in block I was only 226g, but the total kernel weight is used to determine the average number of kernels per cob, and it is from this number of kernels per cob that we assess the yield, and the yield for T14 was only 1.45T/Ha.However, in the other two blocks, Block II and Block III, the yield obtained by T14 is quite good compared with all the treatments, at 3.2T/Ha and 2.7T/Ha respectively. Pest problems have a negative impact on yield (ANDRIAMAMPANDRY., 2010).

The first hypothesis, stipulating that among the treatments tested there is a combination of giving a very high return compared with others **is verified**.
In the plots treated without nitrogen but combined with calcite (T2, T8), yields were good, unlike when TSP was used. This can be explained by the fact that calcite contains phosphorus, humus and nitrogen (see Appendix II).

Comparing T1 and T2 allows us to determine the effects of calcite and TSP during the experiment. These two treatments differ only in the type of phosphate fertiliser used.With T1 (C0N0P1), we obtained a yield of less than 3T/Ha, whereas with T2 (C0N0P2) the yield was 4T/Ha. Calcite had a positive effect on increasing grain maize yields. Calcite is a natural phosphate in tricalcium form derived from micronised apatite rock and volcanic black earth rich in humus and micro-elements (PROCHIMAD). Humus is a source and reserve of plant nutrients (CIRAD and GRET, 2006). Under the action of soil microbes, humus gradually mineralises, releasing not only nitric nitrogen, but all the fertilising elements and trace elements that were integrated into the organic matter (GROS A, 1962). Humus also promotes the action of mineral fertilisers.As well as humus and nitrogen, calcite also contains 40% lime (Ca0), which is a limestone amendment. Soil improvers are products that improve the overall physical, biological and chemical properties of the soil. Ca0 is also a plant food that promotes growth, gives resistance to plant tissue and influences seed formation and maturation (GROS A., 1962).

The use of calcite has therefore made up for the lack of nitrogen in the plots. It helps to improve the physical, chemical and biological properties of the soil. In addition, natural phosphorus has a residual effect, helping to increase yields for the following year's crops.The results of the experiment show that TSP is very interesting if it is combined with nitrogen at a dose of between 60Kg/Ha and 120Kg/Ha, the experiment on TSP and urea. by RASOAMAHARO (2008) confirms this result. Crushed hyperphos phosphate (calcite) is the most interesting, even if it is not combined with nitrogen fertilisers. However, the study as a whole shows that the highest yields were obtained using TSP.

The **second hypothesis**, that calcite has a positive effect on yield and soil properties compared with TSP, has been **partially verified.**

The treatment C N P_{222} was eliminated, logically, treatments T10 (C N P_{112}) and T12 (C N P_{122}) should be part of group A having given a high yield, but they are not part of it, nevertheless the conditions to have a good production were satisfied for these two treatments. This can be explained by the fact that these two treatments with the T14 treatment in block I are located side by side and there was an attack of earworm and cutworms. Part of block I is marked by the presence of these insects despite the insecticide treatments carried out for the whole of the experimental field.

Helicoverpa zea is a species of lepidopteran insect in the Noctuidae family, native to North America. The caterpillar, or "corn earworm", feeds on and burrows into the ears of maize, making it one of the most destructive pests of maize crops. The caterpillar consumes the silks, preventing fertilisation. This has had a direct effect on grain maize yields. In addition, these caterpillars penetrate the ear tip and devour the tender kernels (FARE Y, 2004). In general, the actions of phosphorus, nitrogen and potassium are complementary for the plant. Phosphorus plays a major role in fertilisation and grain formation, as well as root development (RASOAMAHARO, 2008). It helps the plant resist drought and improves nutrient uptake (RANDRIAMANANANDRO., 2010), while potassium leads to ear and grain formation. As for nitrogen, as various authors have pointed out, it is the linchpin of fertilisation for high yields.

2- Effect of the factor studied on yield components

With regard to the number of grains per ear, of the 18 treatments, 11 gave a higher number of grains per ear than the 7 remaining treatments. These were : T11, T17, T9, T3, T5, T15, T6, T12, T16, T2 and T8. The study of the interaction between the number of grains per ear and nitrogen as well as between the number of grains per ear, nitrogen and phosphorus had a significant effect. The number of grains per ear therefore varied as a function of the nitrogen dose on the one hand, and varied as a function of the combination of phosphorus and nitrogen on the other. Most of the plots with these treatments received nitrogen, which plays a key role in increasing the number of grains per ear.the number of grains per ear (ANDRIAMAMPANDRY, 1990). A good supply of water and nutrients (nitrogen in particular), especially at the critical moment of flowering and ear formation, will limit seed abortion and give a long, well-filled ear

(GROS A, 1962). T2 and T8 are treatments with no nitrogen input but treated with calcite containing nitrogen.The number of spikes per plant was more or less the same for all the plots in the three blocks, although some treatments, such as T9, had more than one spike per plant. For 1000-grain weight, it did not vary according to the factor studied (nitrogen-coverage-phosphorus) or with one of the components of the factor studied (nitrogen, phosphorus, cover, nitrogen and phosphorus, nitrogen and cover, phosphorus and cover). This can be explained by the fact that 1000-grain weight is a fixed varietal characteristic, so there was no significant difference between 1000-grain weights and treatments. Nitrogen is the element that interacts with these yield components. All this shows that nitrogen is the main factor in yield in terms of fertilisation. Without nitrogen, the weight or number of yield components in grain maize decreases.

3- Lift rate

The emergence rate was 55% for all the experimental plots. Treatment effects were noted, but the block factor had no influence on the emergence rate. There were plots with a high emergence rate and plots with a low emergence rate. We found that the emergence rate for plots treated with a nitrogen dose of 120Kg per hectare was very low, varying from 35 to 50%. Plots treated with N0, i.e. no nitrogen, had the highest emergence rates, up to 80%.The urea dose of 120Kg/Ha is one of the causes of the low seedling emergence rate. The urea used in the experiment had a negative impact on seed emergence. Contact with the urea caused the seeds to burn. RASOAMAHARO (2008) confirms this burn caused by high-dose urea in his final dissertation. The excessive concentration of nitrogen on the seed can burn it (GROS A, 1962). If urea is applied locally during sowing, the maximum rate of nitrogen applied at sowing for localised maize should not exceed 15-20 kg/Ha (GROS A, 1962). In our experiment, the maximum dose of nitrogen applied during sowing was 40Kg/Ha, far exceeding the

dose recommended by GROS A. In addition, there was no rain in the study site the day after sowing, the day after sowing and the two days before sowing, and localised nitrogen should be avoided, especially in the event of prolonged drought (GROS A, 1962).Apart from the dose of urea on the one hand, there's also the rainfall on the other.
Water, light and temperature are external factors in the germination of maize seeds (FARE.Y, 2004).Maize grain can only germinate if the temperature is at least 10°C (zero germination). At this temperature, emergence will take 15 to 20 days. At 20°C, emergence will only take 8 to 10 days to be complete. It is preferable to delay sowing rather than plant the crop in poor conditions (too cold or too wet), to ensure a good start to the crop and good vigour at the start (AB technical data sheets, 2012). This temperature condition was met for our experiment because the temperature between sowing and emergence was at least 14.5°C (Ampandrianomby weather station, 2014) and emergence took place around D_{10} after sowing. However, after sowing, the soil was too wet due to heavy rainfall at the study site. A few days after sowing the maize, rainfall was very high in the Vakinankaratra region. At the time of sowing on 30 November, rainfall was 0.5 mm, and after a few days on 3 and 05 December, 16.6 mm and 28.8 mm respectively, as shown in the graph below. Plots in blocks I and III were flooded. About 10% of the maize seeds in some plots rotted. This is why we had to eliminate the T18 treatment because the rate of emerged plants on the plots treated with T18 in blocks I and III varied from 20-25%.

In terms of emergence rate per block, block II had a higher emergence rate than the other two blocks. However, the difference in emergence rate between the three blocks is not due to the factor studied. During the observations, we noticed that block I and block III were very flooded compared to block II, which explains why the lift rate in block II is higher than that in block I and block III. This difference in emergence rate between blocks is also due to a number of factors such as soil structure, seed placement linked to tillage and sowing operations (BRUNEL S, 2010 and RAJAONARIVELO S, 2012).

Figure 15: Daily rainfall in the study site

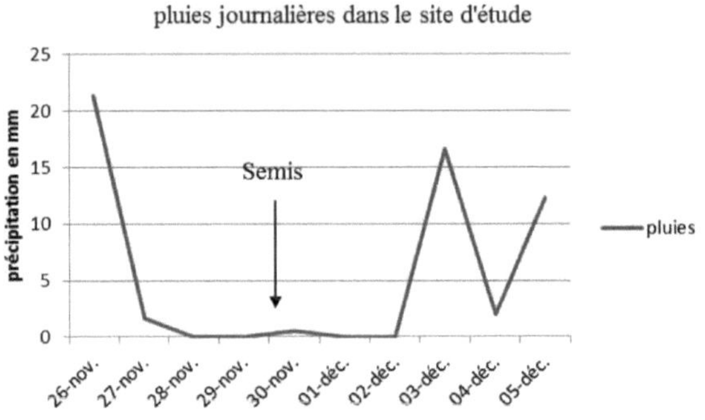

pluies journalières dans le site d'étude

précipitation en mm

25
20
15
10
5
0

Semis

pluies

26-nov. 27-nov. 28-nov. 29-nov. 30-nov. 01-déc. 02-déc. 03-déc. 04-déc. 05-déc.

Source: author

4- Flowering and heading

Maize is a monoecious plant, with a male inflorescence and separate female inflorescences on the same maize plant. In maize, the male inflorescence appears when the male flowers and the female inflorescence when the female flowers. The male inflorescence appears before the female inflorescence.The male inflorescence is a terminal panicle made up of spikelets each containing 2 male flowers. Female inflorescences number 1 to 4 per plant. They are located in the axils of the leaves in the middle of the stem. They are spikes wrapped in rudimentary leaves called "Spathes". Each spike consists of a "rafle" on which are inserted in vertical rows hundreds of spikelets with 2 female flowers, only one of which is fertile. At the moment of fertilisation, the flower styles emerge from the end of the spikes in the form of green or pinkish bristles. In the experiment, the male inflorescence appeared between D_{76} and D_{86} after sowing, and heading occurred around D_{93} . There was no significant difference between the date of flowering and heading for each of the treatments.

5- Maturation

During the observations, the plots with treatments T1, T2, T7, T8, T13 and T14 matured before the other plots in the three blocks. The silks and panicles of the maize in these plots dried out before those in the other plots. What T1, T2, T7, T8, T13 and T14 is the fact that the nitrogen dose used is N0, i.e. no nitrogen was added. On the other hand, the plots treated with N2, i.e. with a nitrogen dose of 120Kg/Ha, had a delayed ripening compared with N1 and N0. This is explained by the fact that nitrogen delays plant senescence and maturation (ANDRIAMAMPANDRY, 1989 and MABA, 2007).

6- Height

6- 1 Height to J_{30} AS:

The treatment as a whole had no significant effect on plant height. However, the phosphorus factor did have an effect and we carried out a pairwise comparison of the effect of phosphorus on plant height. Like nitrogen, phosphorus is a plant growth factor (GROS A, 1962 and RAKOTOARISOA, 2009). During the first growth stage, the plant has very high phosphorus requirements, which are covered by seed reserves. It is mainly absorbed during the active growth period (RAKOTOARISOA, 2009). Once these reserves are exhausted, the young plant quickly shows signs of deficiency.

6- 2 Height at J_{56} AS:

Although the treatment as a whole did not have a significant effect on height, nitrogen and phosphorus had an effect on height growth. For nitrogen, around 1 month after the 2/3 urea was applied, there was an increase of height of IRAT 200 maize.This contribution encouraged the plants to grow in height. Growth corresponds to the multiplication of cells from the first cell division and the increase in cell size. Water, light, temperature and mineral elements are the limiting factors in this growth, apart from external factors. Nitrogen was a limiting factor in our experiment. A plant that is well supplied with this element grows quickly and produces a lot of leaves and stalks (GROS A 1962 and FAO, 2002).With regard to N0, a nitrogen deficit (according to Morgan, 1990 and FAO, 2002) encourages an increase in abscisic acid in the plant, and this acid is a general inhibitor of cell growth, which affects the height growth of maize. As we said earlier, phosphorus, like nitrogen, is also a growth factor for plants. Its action is complementary to that of nitrogen (RAKOTOARISOA., 2009).

6-3 Height at the end of the cycle

As the interaction between height and treatment did not have a significant effect overall, we carried out a pairwise comparison for the factors that did have an effect. For final height, the use of phosphorus had a significant effect. This reinforces the theory that phosphorus is a growth factor for plants.With regard to the effects of phosphorus, a pair-wise comparison of the average heights obtained by P1 and P2 showed that the plots treated with P2 (calcite) had a lower height than the plots treated with TSP (P1). This difference in height is explained by the fact that TSP is a fast-acting phosphate fertiliser that acts mainly during the plant's growth phase, whereas calcite is a slow-acting fertiliser. The slow-acting calcite (PROCHIMAD) is still available to the plant for seed formation and maturation. In terms of height and yield, plant height growth had no influence on grain maize yield. This explains why, even though the height of plots treated with calcite was low, their grain maize yield remained high.

7- Rainfall

The relationship between rainfall and maize cultivation was observed during this experiment. The complete cycle of IRAT 200 during the experiment was approximately 123 days. Rainfall data for Antsirabe I were collected at the Ampandrianomby weather station.The germination phase, from sowing to emergence, lasted 15 days in our case. To germinate, the grain needs water. The soil must have acquired a reserve even before sowing; two to three days of rain can guarantee good emergence (FARE Y, 2004). This need for water before sowing was met, as shown in the figure below:

Figure 16: Daily rainfall in Antsirabe from 26 November to 30 November 2013 Three days before sowing and on the afternoon of sowing itself, it was raining in the study area.

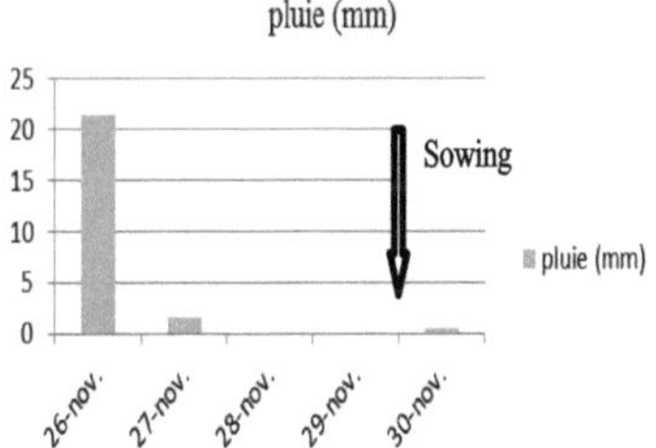

Source: author

After the germination phase, maize enters the growth phase. Maize grows slowly. In this study, the growth phase began around D_{15} after sowing and ended around D_{76} after sowing. During the growth phase, which is the stage between emergence and the appearance of the panicle, IRAT's water requirements (around 100 mm per month) were met, as shown in the figure below:

Figure 17: Monthly rainfall from 15 December 2014 to 15 February 2014

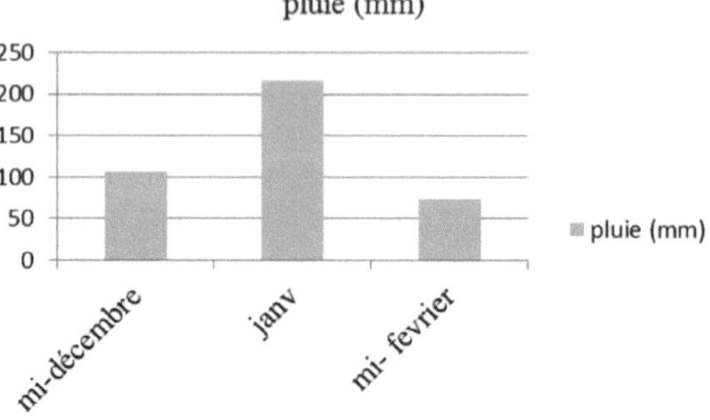

Source: author

Once growth is complete, the male inflorescence appears. In our case, this occurred around D_{76} after sowing, on 15 February. Water requirements are much higher for flowering. A week after the appearance of the male flowers, the female flowers appear.The critical water period for maize crops is 15 days before the appearance of male flowers and 15 days after (GROS A., 1962, FARE Y., 2004). This water requirement was met in our experiment, as shown in the figure below:

Figure 18: Monthly rainfall during flowering and heading

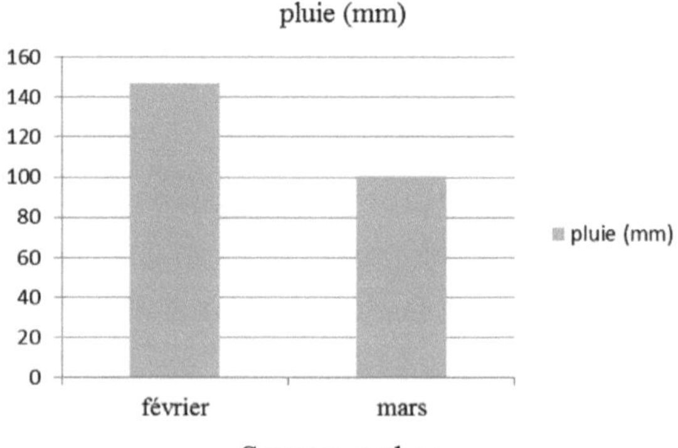

Source: author

The fertilisation phase is extremely dependent on water and nutrients. For ripening, dry weather is favourable, with 39 mm of rainfall in April.

Figure 19: Monthly rainfall during fertilisation and ripening

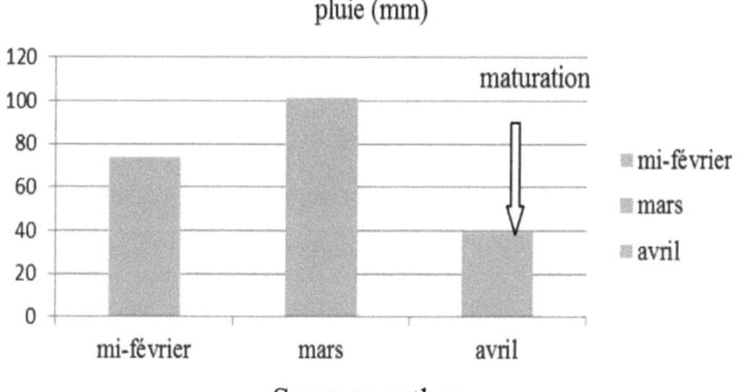

Source: author

The water requirements of maize during our experiment were met. As mentioned above, maize's critical water requirements are 15 days before and 15 days after the appearance of inflorescences. A lack of water during this phase can cause the panicle and silks to dry out, resulting in poor fertilisation, which reduces the number of grains and lowers yields (MUHAMMAD B, 2001 and FARE Y, 2004). Water was therefore not a limiting factor in yield, although it did have a negative impact by flooding some plots.

IV- RECOMMENDATIONS

Recommendations are proposed to improve the research, ensure the reliability of the results and ensure the widespread use of effective treatments during this experiment.

To confirm the results obtained and the effectiveness of the treatments, the experiment should be repeated in time and space. The use of control plots (a practice identical to that used by farmers) is recommended in order to compare the results obtained with the treatments and those obtained with the controls.To obtain a high emergence rate, it is advisable to localise the nitrogen but in small quantities.

This means that it is very useful to divide up the nitrogen two or three times throughout the crop cycle. The first application at sowing, the second 45 days after sowing and the third around the second half of the male flowering period (MAEP, maize data sheet).

To get the most out of nitrogen fertiliser and achieve higher yields, nitrogen must be applied when the plant needs it most. In fact, nitrogen fertiliser is a precision fertiliser that will work all the better if it is available to the plant when it needs it most. Hence the usefulness of applying nitrogen when the male flowers appear, given that a large proportion of maize's nitrogen requirements are met 15 days before and after the appearance of the male inflorescences.

Choosing the right sowing date is crucial to getting the seedlings off to a good start and ensuring the success of the experiment. The use of ground hyperphos phosphate (calcite), a natural phosphorus, is proving useful, and is especially to be recommended for perennial crops and for organic farming whose aim is to have sufficient production of high nutritional quality, and to maintain and improve soil fertility in the long term. Indeed, the composition of calcite is very interesting for the objectives of organic farming. Even when not combined with nitrogen and blankets, calcite gives a high yieldof grain maize, it is very effective. Lastly, as we need to find the right fertilisation to achieve high yields, even if yields are already high with the fertilisers used, the use of organic manure would be ideal.

CONCLUSION

Maize is a very important crop both globally and in Madagascar, and is the subject of a great deal of research and experimentation. Research has been carried out by AGRIVET on the fertilisation of IRAT 200 maize in the Vakinankaratra region, one of the maize-producing areas. 18 treatments with different combinations and doses of phosphate and nitrogen fertilisers with plant cover were tested. The interaction between the 18 treatments and maize grain yield had a significant effect. During experimentation, the best yields obtained were from treatments using TSP. Among the 18 treatments, we found a group with the highest yields between 4T and 5T/Ha compared with the other treatments. This group is made up of treatments T9 (C1 + N1 + P1), T11 (C1 + N2 + P1) and T17 (C2 + N2 + P1). Of the different combinations and doses of fertiliser with cover, T9 is the treatment with the highest yield. TSP is therefore a very useful phosphate fertiliser for high yields. As for calcite, if nitrogen fertiliser is not used, this natural phosphorus is recommended for high production.The results of this experiment show that an increasing dose of nitrogen increases maize yields, especially if it is applied in the right proportions. Of course, to ensure the reliability of the results, repetitions in the coming year are useful, as is the use of a control plot managed in the farmers' own way. So the question is: "Is the TSP treatment with nitrogen, which gave a high yield in this experiment, interesting from an agronomic point of view and economically profitable compared with the control plot?

BIBLIOGRAPHY

ANDRIAMAMPANDRY Dieu Donné. 1990. Le facteur fertilisation dans le développement maïsicole de la région d'Itasy. Dissertation for the diploma of agricultural engineer, ESSA, Agriculture Department. 68p.

BADJECK B and NDIAYE Cheikh Ibrahima (FAO) and Francesco SLAVIERO (WFP), 9 October 2013. FAO/WFP food security assessment mission to Madagascar.

BAMBARA F., Juin 2012, Optimisation de la fertilisation azotée du maize en culture pluviale dans l'ouest du BURKINA FASO.60p.

BRUNEL S., 2010. Ecophysiological characterisation of different Medicago Truncatula genotypes during the germination and heterotrophic growth phases. Doctoral thesis in Agronomic Sciences. 105p.

CANTIN Jean. , 2003. Effects of phosphate mineral fertiliser inputs to grain maize starters to supplement phosphorus inputs from farmyard manure according to soil phosphorus saturation.

CHAMBER OF AGRICULTURE of New Caledonia, 2000. Fiches techniques des engrais de la chambre d'agriculture. 23p.

CIRAD-GRET, 2006. Mémento de l'agronome. Ed. Du GRET, ed. Du CIRAD; French Ministry of Foreign Affairs.

DRILLAUD Céline, 2013. Performance of nitrogen fertilizers on wheat and maize. 46p.

Entreprise DUPONT, June 2011 Nutrient deficiencies in maize and soybean volume 21 number 2. 4p.

FAO, 2002. Maize in the tropics: improvement and production. 382p.

FARE Y., 2004. Experimentation agronomiques et compréhension des systèmes de production paysanne en vue du développement de la culture de maïs dans la région d'Ambohidratrimo. Dissertation for the diploma of agricultural engineer, ESSA, Agriculture Department. 92p.

GROS A.1962. Engrais : guide pratique de la fertilisation, 441p

INRA (French National Institute for Agricultural Research), 2011. Fiche technique : maïs associé aux légumineuses.10p.

LES FICHES TECHNIQUES AB - v. 2012 Groupe technique AB Franche Comté le maïs

LIMAGRAIN, 2010. Maize and water. 6p.

MABA B., 2007. Identification of major limiting nutrients and appropriate fertilisation strategies under maize cultivation in the Eastern Ogou of the Plateaux region. University of Lomé - TOGO

MAEP UPDR - OCEAN CONSULTANT, 2004. Filière maïs fiche n°108. 10p.

MINAGRI et al. 2010. Catalogue national des espèces et variétés cultivées à Madagascar. First edition. 117p.

MUHAMMAD B., 2001.Effect of water stress on growth and yield components of maize variety YHS202.University College of Agriculture. Pakistan.

NDIAYE and SIDIBE, 1992. Recherche de formules d'engrais N-P-K économiquement profitables pour la culture du maïs pluvial au Sénégal. 29p

NOURA Ziadi, 2007. Utilisation des engrais minéraux azotés en grandes cultures : description des différentes formes et leurs impacts en agroenvironnement. 29p.

PROCHIMAD. Technical data sheet on ground hyperphos phosphate, biological product. 2p.

RAJAONARIVELO Sarah Y., 2012. Performance tests of PANNAR maize hybrid varieties in the Vakinankaratra and Bongolava regions. Dissertation for the diploma of agricultural engineer, ESSA, Department of Agriculture. 44p.

RAKOTOARISOA Tiaray H., 2009. Analyse des composantes de rendement des variétés de riz pluvial NERICA sous diverses sources d'engrais phosphatés. Dissertation for the diploma of agricultural engineer, ESSA, Agriculture Department. 88p.

RANDRIAMANANANDRO A., 26 January 2010. Effects of TSP and cattle manure on rainfed rice in a rice-rice succession. Case of a ferralsol in Laniera. Final dissertation for the diploma of agricultural engineer, ESSA, Agriculture Department. 33p.

RASOAMAHARO lova. ,2008. Effect of guano and TSP on maize cultivation: case of a ferralsol in Lazaina Antananarivo. Dissertation for the diploma of agricultural engineer, ESSA, Agriculture Department. 54p.

SERVICE DE LA STATISTIQUE AGRICOLE (INSTAT), 2012. National maize production.

THIBAUDEAU S., Mai 2006, Agriculture, pêcheries et alimentation au Québec. Fertilisation azotée dans le maïs grain. 8p.

RURAL DEVELOPMENT POLICY UNIT (UPDR), monograph on the Vakinankaratra region, June 2003

ZAFINDRABENJA A., 2012. Agronomic experimentation on the fertilisation of the onion crop "allium cepa "with guanobarren, Cas d'Anevoka. Dissertation for the diploma of agricultural engineer, ESSA, Department of Agriculture. 29p.

WEBOGRAPHY

Agpm.com, 2012

CIC. Com, 2014

APPENDICES

Appendix I: Maize Maize: Zea mays

1- Classification of maize according to Cronquist (1981)

Kingdom Plantae

Subregion Tracheobionta

Division Magnoliophyta

Class Liliopsida

Subclass Commelinidae

Order Cyperales

Family Poaceae

Subfamily Panicoideae

Tribe Maydeae

Genus Zea

Species Zea mays

Malagasy name: Katsaka

2- Aims of culture (Memento., 2006)

Maize grains are used: for human consumption: green (cooked, grilled, in salads, soups, etc.), dried, as popcorn, for animal feed: used as grain and feed, or the whole plant, harvested when the cob is doughy, is used as fresh fodder or silage.

3- Description (FARE Y 2004, MAEP 2004)

The Roots :

They are of the fasciculated type. They are superficial as they do not exceed 50 cm in depth.

depth. Aerial adventitious roots form on the nodes at the base of the stems.

Les Tiges :

There is therefore a single round stem, more or less grooved, made up of nodes and internodes. The internodes at the base are shorter. The stem is filled with a sweet pith. It is 1.5 to 3.5 m high and 5 to 6 cm in diameter.

The Leaves

They are attached to the stem at the nodes. They consist of a sheath and a leaf blade. flat, a small ligule can be seen between the blade and the sheath. There are no auricles. Inflorescences A male inflorescence and separate female inflorescences are found on the same plant. The male inflorescence is a terminal panicle made up of spikelets each containing 2 male flowers. There are 1 to 4 female flowers per plant. They are located in the axils of the leaves in the middle of the stem. They are spikes wrapped in rudimentary leaves called "Spathes". Each spike consists of a "rafle" on which are inserted in vertical rows hundreds of spikelets with 2 female flowers, only one of which is fertile. At the moment of fertilisation, the styles of the flowers emerge at the end of the spikes in the form of green or pinkish bristles.

Flowers:

The male flowers are made up of glumes and glumellae surrounding 3 stamens. The female flowers each have 1 ovary topped by a very long style. The male flowers bloom before the female flowers. Fertilisation is therefore cross-pollinated.

The fruit

It is a caryopsis. Each grain is arranged in vertical rows (8 to 20 depending on the variety) along the stalk of the ear. The shape of the kernels (globular, ovoid, prismatic, etc....), their colour (white, reddish yellow, golden, purple, black), their size and their type (smooth or wrinkled) vary greatly from one variety to another. Good kernels for choosing seeds are in the middle of the spike, with small ones at the ends. Each grain is made up of a husk, an albumen, a cotyledon and an embryo. There are 500 to 1000 grains per ear. An ear weighs 150g on average.

4- Physiology (FARE Y., 2004)

According to the phenological stages of maize, we have :

- the germination phase

This phase is marked by the swelling of the grain under the influence of moisture. After 2 to 3 days from sowing, the radicle appears.
The stalk appears on the third or fourth day after sowing. This means that maize emerges 8 to 10 days after sowing.

- the growth phase

This phase is very slow from emergence to the appearance of male inflorescences. This stage lasts for varying lengths of time, depending on the variety, temperature and moisture content of the soil. Maize reaches a height of 10 to 15 cm after 4 to 5 weeks from sowing; 2 months after sowing it reaches a height of around 50 to 60 cm.

- the flowering phase

As soon as growth is complete, the male inflorescence appears. This occurs around 70 to 90 days after sowing. 5 to 8 days after the male inflorescences appear, the female inflorescences are ready for fertilisation.

- the fertilisation phase

After 5 to 10 days from the appearance of the female inflorescence, the maize enters the phase of fertilisation.

- the maturation phase

During this phase, once the grains have formed, they pass through three successive stages: the milky stage, the pasty stage and the vitreous stage.

5- Ecology (Mémento., 2006)

Heat requirements For germination, maize needs at least 10°C. During its growth, maize needs an optimum temperature of 19°C.

Water requirements

It is estimated that an average of 100mm of water per month is needed throughout the growing season, as maize is a water-demanding plant, especially during the germination, growth, flowering, fertilisation and grain enlargement phases. But the most critical period for water is the 15 days before and 15 days after the appearance of male inflorescences.

Lighting requirements

Maize needs plenty of sunshine

Altitude requirements

Maize grows both by the sea and on the high plateaux when the above ecological conditions are satisfactory. However, it cannot grow above 1800m altitude.

Soil requirements

Maize is a demanding plant, so the best soils are : Deep, loose, fresh, fairly light, fertile, humus-bearing to avoid the risk of compaction and permanent waterlogging that asphyxiates the roots. Especially alluvial soils from baiboho or from recent volcanism, which contain mineral elements and organic matter.

-slopes < 12% to avoid the risk of erosion

- soils not too acidic, pH< 5

Appendix II: Correlation test and ANOVA

Lift rate

Analysis of variance

Source DDL

Model	18	11270,967	626,165	7,169		< 0,0001
Error	29	2532,823	87,339			
Corrected total	47	13803,790				
Source			author			
Analysis Type I Sum of Squares :						
Sum of Source DDL squares			Average square	F		Pr > F
Treatment 16 11217,802			701,113		8,028	< 0,0001
Block 2 53,165			26,583		0,304	0,740

Sum of squares

Average

Squares F Pr > F

Source: author

Height and yield

Type I Sum of Squares analysis: height - yield

Source	DDL	Sum of squares	Average of squares	F	Pr > F
height 3	1	1,450	1,450	2,566	0,130

Source: author

Height 1 and treatment

Phosphorus / Fisher (LSD) / Analysis of differences between modalities with a 95% confidence interval :

Contrast	Difference	Standardised difference	Critical value	Pr > Diff	Significant
P1 vs P2	5,562	2,777	2,045	0,010	Yes

Source: author

Height 2 and treatment

Nitrogen / Fisher (LSD) / Analysis of differences between modalities with a 95% confidence interval

ModalityEstimated mean Groups

N2	106,485	A	
N1	94,083		B
N0	90,444		B
		Source: author	

Height 3 and treatment

Phosphorus / Fisher (LSD) / Analysis of differences between modalities with a 95% confidence interval :

Contrast	Difference	Standardised difference	Critical value	Pr > Diff	Significant
P1 vs P2	13,359	3,384	2,045	0,002	Yes

Source: author

Treatment and performance components

- Number of grains per ear

Analysis of variance: number of grains per ear and treatment

Source	DDL	Sum of squares	Average square	F	Pr > F
Model	18	237113,670	13172,982	4,715	0,000
Error	29	81014,131	2793,591		
Corrected total	47	318127,801			

Source: author

Analysis of ISS type: number of grains per ear and treatment

Source	DDL	Sum of squares	Average squares	F	Pr > F
Cover	2	4302,076	2151,038	0,770	0,472
Nitrogen	2	105634,367	52817,184	18,907	< 0,0001
Phosphorus	1	873,895	873,895	0,313	0,580
Block	2	19805,221	9902,611	3,545	0,042
blanket*nitrogen	4	5873,298	1468,324	0,526	0,718
cover*phosphorus	2	11959,091	5979,545	2,140	0,136
nitrogen*phosphorus	2	23172,357	11586,179	4,147	0,026
cover*nitrogen*phosphorus	3	65493,364	21831,121	7,815	0,001

Source: author

Number of ears per plant :

Analysis of variance :

Source	DDL	Sum of squares	from Average squares	from F	Pr > F
Model	18	0,323	0,018	0,945	0,539
Error	29	0,551	0,019		
Corrected total	47	0,874			
			Source: author		

Analysis Type I Sum of Squares :

Source	DDL	Sum squares	from	Average squares	from	F	Pr > F
Cover	2	0,025		0,012		0,655	0,527
Nitrogen	2	0,007		0,004		0,188	0,829
Phosphorus	1	0,003		0,003		0,166	0,687
Block	2	0,061		0,031		1,608	0,218
blanket*nitrogen	4	0,082		0,020		1,078	0,385
cover*phosphorus	2	0,019		0,010		0,506	0,608
nitrogen*phosphorus	2	0,010		0,005		0,257	0,775
cover*nitrogen*phosphorus	3	0,116		0,039		2,034	0,131

Source: author

Weight of 1000 grains

Analysis of variance :

Source	DDL	Sum of squares	fro m	Average squares	fro m	F	Pr > F
Model	18	135,702	7,539		0,933		0,550
Error	29	234,215	8,076				
Corrected total	47	369,917					
			Source: author				

Analysis Type I Sum of Squares :

Source	DD L	Sum of squares	fro m	Average squares	fro m	F	Pr > F
Cover	2	1,144		0,572		0,071	0,932
Nitrogen	2	4,140		2,070		0,256	0,776
Phosphorus	1	0,720		0,720		0,089	0,767
Block	2	41,743		20,871		2,584	0,093
blanket*nitrogen	4	26,255		6,564		0,813	0,527
cover*phosphorus	2	43,725		21,863		2,707	0,084
nitrogen*phosphorus	2	3,929		1,964		0,243	0,786
cover*nitrogen*phosph or							
e	3	14,046		4,682		0,580	0,633

Source: author.

Performance and treatment

Analysis of variance performance and treatment

Source	DDL	Sum of squares	Average squares	F	Pr > F
Model	18	31,275	1,737	2,306	0,022
Error	29	21,851	0,753		
Corrected total	47	53,125			

Source: author

Analysis Type I Sum of Squares :

Cover	2	0,481	0,240	0,319	0,729
Nitrogen	2	7,340	3,670	4,871	0,015
Phosphorus	1	0,006	0,006	0,008	0,929
Block	2	3,241	1,621	2,151	0,135
blanket*nitrogen	4	3,848	0,962	1,277	0,302

Source DDL

Sum of squares

Average

Squares F Pr > F

cover*phosphorus	2	3,548	1,774	2,355	0,113
nitrogen*phosphorus	2	3,705	1,852	2,459	0,103
cover*nitrogen*phosphorus	3	9,104	3,035	4,028	0,016

Source: author

Appendix III: The chemical composition of HYPERPHOS BROYE H-B calcite contains

$P\ O_{25} > 20\ \%\ CaO = 40\ \%$

$Total\ N > 0.02\%\ K\ O_2 = 0.3\%\ SiO_2 = 0.5\%\ MgO = 0,3\ \%$

Humus > 1.0

Rich in micro-elements such as iron, boron, copper, molybdenum, zinc, etc.

Annex IV: Composition of the TSP

Chemical formula: $Ca(H_2 PO_4)2H_2 0$. TSP contains approximately 48.7% P 0_{25}

Nom anglais Autre nom français	Triple Superphosphate (TSP) Phosphate monocalcique (composant essentiel)			
Provenance	Sico (Belgique) et Bush Int. (NZ).			
Formule chimique	$Ca(H_2PO_4)_2, H_2O$			
Analyse	N	P_2O_5 (P)	K_2O	CaO
	0	46 % (20 %) (91% soluble dans l' eau)	0	15 %
	Autres : S : 1,3 %			
Solubilité	Assez soluble dans l' eau (18 g/L).			
Humidité	6,20 % max.			
Granulométrie	< 2 mm : 1 % 2-4 mm : 85 % > 4 mm : 14 %			
Densité	0,85-0,95 t/m^3			
Hygroscopicité	Très faible (au delà de 95 % d' humidité relative à 30 °C).			
Stockage / précautions	Garder le sac fermé avant utilisation. Stocker en dehors de l' action directe du soleil. Se laver les mains après utilisation.			
Compatibilité	Mélange toujours possible : - ammonitrate, nitrate de potassium, sulfate de potassium, 0-32-16, 13-13-21, 16-4-8, fumier, engrais organiques. Mélange possible au moment de l' emploi : - urée, MKP, 17-17-17, gypse. Ne jamais mélanger avant emploi : - nitrate de calcium, hyperphosphate, chaux.			
Effet sur le pH	Peu d' effet sur le pH du sol.			

Appendix V: Urea (New Caledonia Chamber of Agriculture, 2000)

Urea is a compound in the amide group, and is an ammonium fertiliser containing around 46% nitrogen.

Nom anglais **Autre nom français**	Urea Perlurée (granulométrie surtout de 1-2 mm), Carbamide, Diamide carbonique			N° CAS : 57-13-6
Provenance	Petrochem Ltd. (NZ), ou Sico (Belgique).			
Formule chimique	$H_2N—CO—NH_2$			
Analyse	**N**	**P_2O_5**	**K_2O**	
	46 %	**0**	**0**	
	Autres : -			
Granulométrie	< 1 mm : 3,5 %	1-2 mm : 23,0 %	2-4 mm : 73,0 %	> 4 mm : 0,5 %
Solubilité	Très soluble dans l' eau (1033 g/L).			
Humidité	0,50 % max.			
Densité	0,7-0,8 t/m^3			
Hygroscopicité	Assez forte (dès 72 % d' humidité relative à 30 °C).			
Manipulation / stockage / précautions	Garder le sac fermé avant utilisation. Garder au sec. Stocker en dehors de l' action directe du soleil. Se laver les mains après utilisation.			
Compatibilité	Mélange toujours possible : - nitrate de potassium, sulfate de potassium, fumier, engrais organique. Mélange possible au moment de l' emploi : - superphosphate, 0-32-16, MKP, 17-17-17, 13-13-21, 16-4-8. Ne jamais mélanger avant emploi : - ammonitrate*, nitrate de calcium*, hyperphosphate, gypse, chaux. (*Mélange possible en solution : pulvérisations foliaires et irrigations fertilisantes)			
Effet sur le pH	Tendance à baisser le pH du sol (acidification).			

Appendix VI: Grain maize yield (T/Ha) of treatments

treatment/block	block I	block II	block III
T1	1,80225	3,3665625	3,30075
T2	4,363875	3,1640625	4,647375
T3	2,592	3,7918125	5,153625
T4	3,17925	4,9359375	4,0651875
T5	2,8299375	4,5106875	3,6703125
T6	5,427	4,222125	4,8245625
T7	1,407375	2,541375	2,0908125
T8	4,060125	5,4320625	3,09825
T9	5,2903125	3,7614375	5,93325
T10	3,2754375	3,483	4,8751875
T11	4,96125	3,736125	4,343625
T12	2,754	3,20203125	3,6500625
T13	3,8323125	3,715875	4,039875
T14	1,144125	3,2805	2,69325
T15	3,54375	2,9413125	4,2778125
T16	3,5386875	3,66525	3,331125
T17	5,02453125	5,0979375	4,951125

Source: author

Annex VII: Food production in tonnes in Madagascar (FAO, 2013)

Cultures	Années		
	2012/13 (1)	2011/12 (2)	Ecart (1)/(2) en pour cent
Riz (paddy)	3 610 626	4 550 649	- 21
Maïs	380 848	447 948	- 15
Manioc	3 114 578	3 621 309	- 14

Source: Enquête CFSAM 2013 et StatAgri (MinAgri)

TABLE OF CONTENTS

Printed by Books on Demand GmbH, Norderstedt / Germany